The author has degrees in Physics and Mathematics obtained from the University of Glasgow and Massey University in New Zealand. His PhD degree is in particle and nuclear physics. He has gained qualifications in chemistry, astronomy, geology and philosophy. His interest in climate originated in 1998. He was born in Leicester, England, attended the City of Leicester Boys Grammar School and then went to Glasgow University where he received his honours degree and PhD. He has lectured in Physics at the University of Glasgow and at the University of Canterbury in New Zealand. He has taught science in schools in New Zealand for 37 years before retiring at the age of 72.

To Diana

Bryon Mann

IS CLIMATE CHANGE MAN-MADE?

Copyright © Bryon Mann 2023

The right of Bryon Mann to be identified as author of this work has been asserted by the author in accordance with sections 77 and 78 of the Copyright, Designs and Patents Act 1988.

All rights reserved. No part of this publication may be reproduced, stored in a retrieval system, or transmitted in any form or by any means, electronic, mechanical, photocopying, recording, or otherwise, without the prior permission of the publishers.

Any person who commits any unauthorised act in relation to this publication may be liable to criminal prosecution and civil claims for damages.

The story, the experiences, and the words are the author's alone.

A CIP catalogue record for this title is available from the British Library.

ISBN 9781035807581 (Paperback)
ISBN 9781035807598 (ePub e-book)

www.austinmacauley.com

First Published 2023
Austin Macauley Publishers Ltd®
1 Canada Square
Canary Wharf
London
E14 5AA

Without the encouragement of my friends and relatives, this monograph would not have been written

Table of Content

Preface	11
Chapter One: Evidence Versus Consensus	13
Chapter Two: Historical Evidence	19
Chapter Three: What Makes Up the Earth's Atmosphere?	28
Chapter Four: Infrared Activity	32
Chapter Five: How the Earth's Surface Cools	36
Chapter Six: How Does Earth's Surface Heat Up?	40
Chapter Seven: What Causes Climate Change?	47
Chapter Eight: All About Carbon dioxide	53
Chapter Nine: Speculations and Perceptions	61

Preface

This is a monograph written with pleasure in mind – my pleasure but I hope that it also brings some enjoyment and illumination to those readers who dare to turn its pages. I enjoy writing and I love teaching, both of which are satisfied by this endeavour. Unfortunately, I do not find pleasure or enjoyment in some of the tedious practices associated with producing an educational monograph. So in the following pages, the reader will not be discovering an index, footnotes or references. This is not because I think that these are not important; they are but I am too lazy and old to bother with them. I have spent many years reading the subject matter and find no pleasure in listing everything that I have read and studied. I feel that if my reader is really that interested, then he or she will endeavour to use the internet, which is so available and useful to do their own research. What I do hope is that this monograph will stimulate interest, discussion and debate. It should succeed in doing that since it contains lots of controversial ideas. I don't pretend that these ideas are new but having them in a short and concentrated form will help the reader to formulate an opinion on the subject of whether mankind is responsible for climate change. I do not deny the validity of climate change; I am a great believer that climate change exists.

It is a fact and we can do nothing to alter that. Climate change occurs naturally. This may cause some readers to assume that I also deny that pollution is caused by humans. That assumption is wrong and like many other people, I am very much opposed to pollution and I do my best not to increase the unpleasant occurrence of it. But this monograph is not about pollution and so there is no discussion on that subject included in the following pages. This monograph is entirely about the improbability that global warming is caused by human beings by their emission of carbon dioxide into the atmosphere.

In my last chapter, I have deviated from my stated intention of producing evidence and nothing but evidence. I have presented speculation, rumour and hearsay material in that final chapter because I felt that I would have left those readers, who have managed to digest the first few chapters, with a big question. I, myself, have worried over this question as to why such an obviously wrong myth should be believed by such a large population of the world, including large numbers of influential individuals such as politicians and celebrities. I am not an expert in politics or sociology, so I can only suggest unscientific and evidence-free reasons for this puzzling phenomenon. Perhaps you, dear readers, can throw a more illuminating light on this problem.

Bryon Mann: June 2022

Chapter One
Evidence Versus Consensus

The title of this monograph is in the form of a question which can be classified as a scientific question. It is scientific because the two possible answers are scientific statements; namely "Climate Change is Man-Made" or "Climate Change is Not Man-Made." They are scientific because the validity of either of these statements can be gauged and validated by examining evidence. This is how science works. The scientific method involves collecting and weighing evidence for and against the validity of an hypothesis. A scientific statement must be falsifiable; that means it must allow itself to be tested against evidence obtained by observation and experiment.

Predictions

The hypothesis becomes more acceptable and more useful if it is able to stand the test of making sensible predictions which turn out to be correct. The predictions must be described in a detailed and complete form so that they can be tested by anyone who is able to make the observations or do the experiments needed to test the hypothesis.

There is little use if the prediction relates to something that is impossible to verify; for example, if it relates to an event in the far future, which it so often is for the predictions made by those who maintain that climate change is man-made. We can't test a prediction that states something that is to occur 50 years into the future. It is just as useless if the prediction requires an impossible expenditure of time and effort to test it. If the hypothesis fails to make a satisfactory prediction, then the hypothesis must be discarded and considered as untrue.

It must be emphasised that the only way that one can decide on the truth of the statement "Climate Change is Man-Made" is by subjecting it to the rigours of the scientific method. This means that any predictions made assuming the truth of this statement must be measured against evidence, more evidence, nothing but evidence and only evidence which can be easily collected and weighed. The capable scientist does not consider as relevant evidence the opinions of individuals if they do not attach testable predictions to their opinions, whether they are considered as experts or simply members of the informed lay public. One should endeavour to question even the opinions expressed by a consensus of experts. Certainly, the opinions of the general public should not be accepted as truth without detailed and critical examination. There are just too many examples from history where consensus and authority have subsequently been shown to be wrong.

Expert Opinions Fail

One example of a popular belief that failed was held not only by the general population but also by the authorities and so called experts of that time and that is concerning the motion of the earth. The consensus view since humankind first thought about it was that the sun moved across the sky and rotated around the earth. It was believed that the earth was the centre of the universe. It was so firmly fixed in the general mind set, it was dangerous to question the truth of this doctrine, especially since this idea was upheld by the all-powerful church of that time. Copernicus was aware of the opposition that he would receive, so he waited until he was on his death bed and only then was he brave enough to publish his treatise throwing doubt on the idea that the sun rotated about the earth and instead proposing that the earth rotated about the sun and was not the centre of the universe. We had to wait for the observational evidence made by Galileo with his telescope before there was a gradual acceptance of the sun-centred universe.

One does not need to go too far back in history to find examples where the accepted theories have been found to be wrong. Alfred Wegener suggested at the beginning of the twentieth-century that the continents of Africa and South America had moved apart. His theory was received with contempt and ridicule from the geological experts of the time. The authoritative view was that this suggestion was nonsense. This opposition and ostracism from the consensus of experts lasted for many years until overwhelming evidence in 1950s made continental drift the basis of modern geology.

Another modern example of the downfall of an accepted theory was the belief that stomach ulcers were caused by stress and eating spicy foods. It was dogma that bacteria could not survive in the stomach because of the acids which are secreted in digestion. It was not until the doctors, Warren and Marshall, presented their evidence some 30 years ago that this particular common belief was destroyed and that the bacterium, helico-pylora-bacter, survived in the acidic digestive juices and was accepted as a major cause of stomach ulcers.

Putting sleeping babies face down in their cots was a procedure taught to new mothers by the experts up to quite recent times. When evidence was presented showing that this approach to sleeping babies was a likely cause of cot deaths, this authorised technique was dropped with no apology from those who thought they knew better.

Pasteur produced overwhelming experimental evidence that brought down the theory that miasmas caused diseases and replaced it with the bacterial origin of infections. Until that time miasmas were believed by experts as well as by the lay public to be the cause of disease.

Darwin expected and received a lot of opposition to his theory of evolution. It is only when his *Origin of the species* was published, listing his mountain of evidence which showed that evolution was the process that is now accepted as the basis of biology.

Opposition to the General View is Dangerous

Sometimes the presenting of theories that oppose the authorised views could be dangerous. You had to be brave or foolish to suggest opinions that were against the powerful authorities. One of the examples that come to mind consist of the burning to death of Giordano Bruno because he claimed that stars were in fact suns, which might have planets rotating about them; an idea very much out of favour with the church authorities but is now generally accepted as a fact.

The story of Galileo who suffered imprisonment and ostracism for publishing his theories about astronomy is well known and that his ideas were only gradually accepted by the authorities in the church and by the general population.

It needed more than the evidence of the telescope to persuade them of the truth.

One example that is not so well known is the story of Ignaz Semmelweis. He was the nineteenth-century doctor in Vienna's General Hospital, who saved the lives of many women who would normally have died from childbed fever. He did this by enforcing the simple practice of the washing of doctors hands with chlorinated lime before examining their women patients. Despite publishing his favourable results, Semmelweis's observations were not accepted because they conflicted with the established opinions of the time. The strength of this opposition was thought to be the cause of his mental breakdown.

And there are many more examples where the accepted and widely accepted views have been demolished by evidence. In fact, a famous philosopher once said that all good theories held today were at one time thought to be rubbish. One may wonder what accepted and world-wide views of the present day will at some time in the future be replaced by something that is at present thought and is declared by the experts to be rubbish.

At least in these modern times and in the western world, one is unlikely to be burnt at the stake for holding opposing ideas. One might expect to lose one's employment and receive unpleasant emails and telephone calls for collecting and describing evidence against the accepted opinion but, hopefully, nothing more dangerous than that.

Barriers to New Ideas

The realisation that the history of knowledge is full of wrong ideas that have been demolished by evidence, produces in me a respect for any person who comes along and presents evidence which attacks the authorised belief. Just expressing an opinion, even if held by a consensus of believers without substantial evidence is not enough to gain my respect. What is even less deserving is when the authorities act in an aggressive manner to block any attempt to question their beliefs.

Unfortunately this appears to be the case with climate change. There are many authoritative opinions that voice the mantra that it is too late for further research and scepticism in dealing with climate change. It is a shame that anyone expressing such scepticism are often described by the majority with derogatory terms such as Deniers, Flat Earthers and Attention Seekers, needing psychological treatment. It is their evidence that should be studied and not their personalities. Science, if it is to make progress, must never stop asking questions and never attempt to eliminate opposing ideas simply because they go against the authorised view.

Chapter Two
Historical Evidence

If this warm period, in which we find ourselves at the present time, was brought about without the aid of industrialisation or any other artefact of man's efforts, then we could rightly state that climate change is natural. If it is natural, then it is quite likely that there would have been other naturally occurring warm periods in the past. If we find that there have been warm periods in the past, then this would throw considerable doubt on the claim that this present warm period is man-made. There was no industrialisation before the nineteenth-century and so man's activities could not be blamed for producing warm periods in the past.

Lack of thermometers

Discovering the existence of past warm periods would require looking for and measuring some sort of permanent record of temperature.

Mankind, in these past times, did not have the intricate and accurate techniques that we have nowadays for measuring temperatures, nor did the philosophers and inhabitants of those past days show any desire or need to measure temperatures. Because of this lack of accurate measurements, it becomes necessary to use our present day technology to ascertain past temperatures by other means.

It is very likely that the people of those times would consider any extreme weather event as an act of retribution of some god or divine creature who requires some sort of sacrifice to placate their anger caused by the sinful act of the populace. Superstition ruled the minds of most people. So it is unlikely that we would expect to find a reliable written record of temperatures. What we often do find are stories and myths, passed down from generation to generation about severe weather events which would have occurred as frequently as they do today. It is only in recent historical times that humans started keeping records of weather in such things as ship's logs and diaries. These, when properly studied, can be used to give some ideas and clues as to the climate of the day.

Proxy Temperatures

Historical evidence is not easy to obtain. Modern scientists have discovered the existence of so called 'proxy temperatures'. These are permanent or near permanent, physical properties which are affected by climate. They include ice cores, tree rings and the abundance of fossil life forms in sediments. These sedimentary organisms thrive well in different climatic conditions, and consequently knowing their abundance and the times when these organisms were deposited, provide scientists with some, and often enough detail to write an approximate history of climate.

There are problems with proxy temperatures. Notwithstanding the difficulty one has of the interpretation of these indirect measurements, one must realise that they are a sort of average and so suffer from the same disadvantages as all averages. Taking averages of a measurement that varies considerably is fraught with difficulties.

Temperatures on the earth are such an example. Temperatures vary considerably from time to time and from place to place. The temperatures during the night are usually cooler than those taken during the day. Temperatures taken during the warmer seasons are usually warmer than those during the cooler seasons. Even from place to place there is a large range of temperatures; the poles of the earth, for instance, are very much cooler than those near to the equator.

This large variation in temperatures makes any average less reliable and should be treated with considerable scepticism.

If one wishes to compare the average temperature of some particular time with that of another time, then, to make it meaningful, one must make sure that each sample is obtained under identical conditions.

Conditions can be subject to subtle changes making measurements less reliable for making general statements. A capable scientist would publish his results with a stated uncertainty, which indicates how confident he or she is in the validity of those results. Unfortunately, those whose job it is to make announcements about world-wide temperatures, hardly ever attach an uncertainty to their published results. Why do they omit this uncertainty? One can only surmise that these uncertainties are so large that they would weaken their argument. A rise in temperature of 1 degree is not dramatic if the uncertainty is plus or minus 2 degrees.

The uncertainty in these averages should be written and attached to the published average. Obtaining an uncertainty on any measurement is usually a simple procedure but with the averaging of temperatures where there is an extensive range of values and a very large field of values from which one takes samples, then obtaining a valid and meaningful uncertainty would require taking lots of similar samples at the same times and at the same places and then analysing them to produce a reliable uncertainty. This is difficult even with modern techniques; applying this to proxy measurements, where one has even less control, is even more hazardous.

Lost Extremes

One other problem with taking averages is the fact that extreme readings tend to get lost in the multitude of less extreme measurements. Extreme measurements do have an effect on the average but just taking the average without an uncertainty, one cannot ascertain whether the distorted result is due to an included extreme value or is due to a slight increase in the value of the multitude of near average measurements. Furthermore, there is a tendency for extreme values, high and low, to cancel each other out, which obviously affects the calculation of the average. These problems are not easy to take into account when finding the average, especially when applied to the very variable temperatures which exist on the earth.

These problems are even more apparent when averaging from proxy measurements. This is because proxy measurements require long periods of time for them to become noticeable. This is because tree rings and other proxy measurements take years and often much longer times, before they are influenced by temperature changes. This means that those extreme values, which only exist for short periods of time, will not have enough time to produce an effect on the average proxy measurement. So the extreme measurements will not appear on the temperature record. Consequently, because of this lack of extreme readings, it is easy to assume that extreme values didn't exist in past climates. This would be an incorrect assumption. Just because the record doesn't show these extreme values does not mean that extreme temperatures did not exist in, for example, the Medieval Warm period or the Roman Warm period or in any other period. There is just no record of them in the proxy readings.

When an extreme weather event occurs in this present day warm period, it is readily measured and reported on, no matter how short a time it lasts. We cannot assume that this is unilaterally a characteristic of the present day climate alone. One must realise that extreme weather events also occurred in past periods; we just don't know about them.

Hot and Cold

Even with the lack of detail obtained from proxy temperatures, scientists have worked out that the earth has been in a dynamic state, temperature wise since it was first formed and have realised that its climate has had a wonderfully varied existence. The climate has ranged from snow ball earths to hot house hells with every climate condition in between. For example, 2.5 million years ago the earth formed ice caps that have not melted away since then. Even so, within that period of time the earth has cycled every 120,000 years or thereabout through extreme ice ages of the length of about 100,000 years and interglacial warmer periods of approximately 20,000 years. The last extreme ice age ended about 15,000 years ago providing us with the present interglacial period of relative warmth.

The fact that these periods are called ice ages and interglacial warm periods does not mean that every day or year in the ice age was cold or that every week in the warm period was perfectly hot. Proxy measurements don't record these short term deviations from the norm.

But scientists have found, superimposed on these ice and warm periods, that there have been climate changes, fluctuating up and down, lasting with time periods measured in centuries. These periods that have lasted for centuries have in hindsight been given names by present day scientists. Scientists like to give names and dates to events, but this often leads to a misunderstanding that there is no variation in that grouping, and that the period had a definite start and finishing date. In reality, it is difficult to state when a warm period starts and a cold period ends. There surely would have been lots of days in between which were difficult to put into hot or cold categories. So the following given times have to be viewed with a considerable flexibility.

This present warm period, for example, is said to have started about 1850. But that doesn't mean the warm period suddenly started with a warm day following a cold day. There were probably lots of days that could easily be placed in either the previous cold period or in the present warm period. There were probably lots of days that were just average. Just because we describe this present period as warm doesn't mean that every day is warm. We still suffer some extremely cold and wet days.

Ignoring the difficulty of being specific about dates, this present warm period could possibly last for another few hundred years if previous warm periods are an indication of their frequency. But it could also last for a much shorter time. We might see the beginning of a new ice age next week. We simply are not able to predict these natural events.

Before this present warm period, there was the Little Ice Age which lasted about 500 years. Stories of the inhabitants of European countries having fetes on the frozen rivers is well known. That cold period started in about 1300 AD, when the Medieval warm period ended. The Medieval warm period lasted for about 400 years. Before, that was the cold period called by present day historians as the Dark Ages, which lasted until about 900 AD. Previous to that there was the Roman warm period (250 BC to 400 AD). And before that the Minoan warm period lasted a thousand years (3000 to 2000 BC).

There was one warm period (about 6000 BC) called the Holocene Optimum which was thought to have been particularly warm. In total, during this 15,000 year interglacial period, there were eight or so previous warm periods.

Medieval Warm Period

During the Medieval warm period, those adventurous travellers, the Vikings, were looking for lands that they could settle in. They wanted land that looked fertile and suitable for farming. They found one that looked particularly green so they called it Greenland. One would unlikely describe it as green at this present time. Because the climate was mild, they built their homes there and lived out their lives there. When they died in their new homeland, they were buried in deep graves, an exercise not so easily done today because of that concrete like permafrost which exists in present day Greenland. When the climate changed to colder temperatures in the onset of the Little Ice Age, these hardy farmers did not survive and the settlements died out.

It has been suggested that another unlikely enterprise taking place in the Medieval warm period, requiring a mild climate, was the growing of grapes in the northern parts of England. There is little doubt that the Medieval Warm period exhibited warm weather conditions that could easily surpass those balmy days of our present time.

Roman Warm Period

It is not surprising to note that citizens alive during the Roman warm period were encouraged by the climate to wear cool clothing. The toga is a particularly appropriate attire for warm weather. It is suggested that even the Roman soldiers, during that warm period, wore uniforms suitable for hot days. Their uniforms had no coverings on their arms and legs because they were not needed in the warm climate. Their footwear consisted of sandals which would be particularly pleasant in hot weather. When the cool weather of the dark ages arrived, their uniforms became more suitable for those lower temperatures, namely sleeves and leggings made their appearance.

It has been suggested that Hannibal, that enemy of Rome during the Punic Wars, was helped by the warm climate to cross the Alps with his army of elephants. Even in today's mildly warm climate, it would not be an easy exercise to travel over the ice covered passes.

No Industrialisation

If these historical warm periods existed in times when there was no industrialisation, then we can definitely and logically state that industrialisation isn't necessary to produce a warm climate. These warm periods could not have been produced by humans emitting carbon dioxide in their burning of fossil fuels because that amount of industrialisation did not occur until the nineteenth century. There is only one conclusion that we can make from the evidence of the existence of past warm periods and that is: climate changes are natural and not man-made.

We must, therefore, ask ourselves the question, "Is our present warm period natural?" If climate change is natural and because there have been so many warm periods in the past, it is not unreasonable to suggest that this latest warm period, in which we now live, is just another warm period in a sequence of naturally occurring warm periods. It would be fantastical and illogical to believe that this latest warm period is any different from previous ones, and, therefore, we should admit that it is not man-made, and is almost certainly quite natural.

Chapter Three
What Makes Up
the Earth's Atmosphere?

The atmosphere of the earth consists of several gases including nitrogen (78% of dry air), oxygen (21% of dry air), argon (rather less than 1% of dry air) and carbon dioxide (0.04% of dry air). These figures have been rounded up to 1% except for that of carbon dioxide. If that figure was rounded to 1%, the carbon dioxide abundance would be zero. Even if the abundances were rounded to 0.1%, then the amount of carbon dioxide in the atmosphere would still record zero percent. The concentration of CO_2 in the air of the atmosphere is remarkably small. Let me illustrate that by an imaginary scenario. If there were no turbulence, no convection currents, no diffusion and no mixing of the gases, then all of the carbon dioxide in the atmosphere would settle close to the earth's surface, being the heaviest of the atmospheric gases. It would just reach above your knees and no further. The rest of atmosphere of oxygen, nitrogen and water vapour would stretch for kilometres above your head.

Water Vapour

The figures quoted in the last paragraph are for dry air but the air in the atmosphere is never dry. This is to be expected since more than 70% of the earth's surface is water, which is constantly resupplying the atmosphere with water vapour. The abundance of water vapour varies considerably but a rough estimate is from 0.4 to 4% of the total. This means that for every molecule of carbon dioxide in the atmosphere there is at least 10 molecules of water vapour and very likely much more.

It has been argued by some climate change advocates that the existence of water in the atmosphere is limited to a certain height because the temperature of the atmosphere at a certain height reaches a low enough value to turn water vapour into a liquid and so falls out of the atmosphere as condensation. So it is argued that above this certain height there is no water vapour and so cannot have any effect on the climate. This argument ignores the fact that supercooled water vapour exists at great heights. Supercooled water vapour is water vapour that exists at very low temperatures without converting to liquid. This unstable form of water vapour will convert into liquid if it is disturbed or if it is able to nucleate on dust particles. A simple observation of this occurring is when a very high flying aircraft produces a vapour trail. The production of a vapour trail can only take place if there is supercooled vapour present. This is good evidence indicating the existence of water in the atmosphere at all heights.

Some other gases in the atmosphere, which are often mentioned and associated with climate change are ozone, oxides of nitrogen and methane but all of these exist in small traces only. Ozone is produced in the upper atmosphere by the reaction of cosmic rays on oxygen.

Ozone protects us from ultraviolet light but it is easily converted back into oxygen and, unfortunately, the conversion is speeded up by the reaction with certain pollutant gases. Even without pollutant gases there is still very little of it to affect the climate.

The oxides of nitrogen are mainly produced by lightning, which produces temperatures high enough to allow oxygen and nitrogen molecules to combine. These oxides are very soluble in water and so are swept out of the atmosphere by condensation. This solution of nitrates is nature's way of fertilising the land. Although its abundance increases over cities, produced by the polluting activities of humans, in general, over the whole earth, the concentration of the oxides of nitrogen are small.

Methane, which is commonly maligned by certain climate advocates because it is highly infrared active and is produced by modern farming is a by-product of rotting vegetation and so tends to accumulate mostly above the equatorial forests. It is gradually oxidised to carbon dioxide and doesn't remain in the atmosphere for very long. The amount of methane in the atmosphere is very small compared with carbon dioxide, which is itself very small; methane's abundance is about 200 times smaller than carbon dioxide, which makes its influence entirely insignificant when dealing with climate change, even if we take into account its great affinity for heat radiation.

Atmospheric Turbulence

The abundance figures quoted above are ratios and not absolute quantities. The atmosphere is in constant motion and so there is plenty of turbulence to produce a good mixing of the gases. The stirring by turbulence is sufficient to make these ratios reasonably constant throughout the depth of the atmosphere. This disagrees with the popular belief that there is a layer of carbon dioxide somewhere in the upper atmosphere. There isn't. There is no evidence that carbon dioxide accumulates in the upper atmosphere or anywhere else.

However, there is a short term and localised concentration over cities but winds and turbulence quickly minimise this aberration. It diffuses into all levels just like all the gases that make up the atmosphere and thereby maintaining its abundance ratio of about 0.04%.

The absolute values of these abundances do not remain static with altitude. This is because the gravitational force of the earth exerted on these molecules is such that the density of the atmosphere decreases rapidly with altitude. The atmosphere is dense near to sea level and becomes very thin at an altitude of a few kilometres. This is what makes it necessary to have pressurised cabins in aeroplanes and the need for oxygen cylinders when climbing high mountains.

This density change is important when studying the temperature at different levels in the atmosphere and is very influential in explaining why it gets colder when we enter these high levels.

Chapter Four
Infrared Activity

Infrared radiation is the scientific name for heat radiation. Infrared activity is a property of all molecules. This includes absorption and emission of heat radiation and is, therefore, very relevant to the global warming argument. Unfortunately, the fact that all molecules absorb and emit infrared radiation, is ignored or denied by many climate scientists. Some molecules, because of their size, are very efficient at absorbing heat radiation and, because of this high efficiency, are consequently called greenhouse gases. It is claimed that these gases act like a greenhouse and are the ones that cause global warming.

Methane, for example, has five atoms in its molecule, so making it into one of the most efficient heat absorbers. But because abundance is also a relevant requirement to the total absorption of heat radiation, methane, with its tiny concentration, adds little to the overall infrared absorption and, consequently, its effect on the climate is negligible.

Both water and carbon dioxide molecules have three atoms in their molecules, so making them next in the ranking of infrared activity. Because water vapour is over ten times more abundant in the atmosphere than carbon dioxide, then this means that its effect on the total absorption of heat radiation is at least ten times greater than carbon dioxide. It is difficult to understand how carbon dioxide is considered by the climate change advocates to be the villain in the climate change argument over and above that of water vapour. It is incredible that many people believe that carbon dioxide has some magical and superior properties that allow it to have an effect on global warming greater than that of water vapour. I certainly do not believe that carbon dioxide has these super properties.

I have searched for references over many years and have not found any which convince me of the existence of these super powers. People simply believe, without any evidence, that carbon dioxide somehow absorbs and stores heat energy greater than does water vapour.

It is often stated that oxygen and nitrogen do not absorb infrared radiation. This is wrong. It is true that these diatomic molecules are poor infrared absorbers and only absorb a small range of wavelengths of the infrared spectrum but the fact that they make up 95 % of the atmosphere means that what they lack in absorption wavelengths, they make up by their overwhelming abundance. The large abundances of oxygen and nitrogen in the atmosphere means that these gases absorb a significant fraction of the total heat absorption in the atmosphere. It is unfortunate that this absorption by the 95 % of the atmosphere is ignored by many scientists.

To understand how these infrared active molecules affect the climate, it is necessary to look at some more science. What happens to these molecules when they absorb infrared energy? The energetic infrared photon is absorbed by the molecule and by so doing, is converted into kinetic energy.

This means that the molecules are energy converters, changing electromagnetic energy of the photon into kinetic energy of the molecule. This extra energy that the molecule receives causes it to move faster or to have stronger vibrations. With this extra energy, the molecule is said to be in an 'excited' state. The excited molecule does not remain in this energetic state for long. It quickly loses its extra kinetic energy in two possible ways.

Sharing Kinetic Energy

It either re-emits the energy as infrared radiation or it transfers some of its kinetic energy to other molecules by colliding with them. This is just like the fast cue ball hitting the other snooker balls, which then scatter in all directions, each of which takes away a portion of the kinetic energy. Similarly in the atmosphere, the energy is shared by further collisions with all the molecules that make up the bulk of the atmosphere. Unlike snooker balls, the gas molecules don't lose their energy by friction; there is simply an increase in the average kinetic energy of the whole gas. This increase in the average kinetic energy is important when discussing the temperature of the gas. This is because the average kinetic energy is another name for temperature.

So this sharing of the excess energy with all the molecules of the atmosphere results in an increase in the temperature of the atmosphere. The atmosphere is big, so reaching this average temperature, over the whole bulk of the atmosphere, takes time, the length of which depends on how massive the atmosphere is. The more massive the atmosphere, the longer it takes to share all of the kinetic energy. The consequential slow warming of the large atmosphere of oxygen and nitrogen effectively insulates the earth and allows life to proliferate comfortably on its surface. This would be difficult without this delay in reaching an overall average temperature. The loss of energy from a less dense atmosphere is faster and by its natural cooling, would lead to an intense cold habitat.

Re-emission of Infrared

The energetic molecule has a second means of losing its energy. Good absorbers of infrared radiation are also good emitters of radiation. This is a well-known principle taught in school physics classes. Unfortunately, re-emission of the infrared is often forgotten, ignored or under-emphasised by many so-called climate scientists, which is sad, because it is very important in the cooling of the atmosphere.

A substantial part of this extra energy is re-emitted as infrared in all directions, not just downwards. Climate scientists seem to emphasise the downward direction of this re-emitted radiation and ignore the upwards directed radiation. The molecules cannot distinguish downwards from upwards. This means that more than 50% of the re-emitted infrared radiation is pointing upwards towards outer space.

Why is it more than 50%? This is because the earth is a sphere and so at any point in the atmosphere, there are more than half of the directions pointing towards the outer perimeter. It is foolish not to take this upward direction infrared into account when discussing the cooling of the earth's atmosphere.

Chapter Five
How the Earth's Surface Cools

It is difficult to understand why there is, amongst the population, the belief that the cooling of the earth's surface is entirely in the form of infrared radiation. This is obviously wrong. The simple experiment of touching a hot surface shows just how nonsensical that is. Heat is transferred by simple contact; this we call conduction. Cooling by emission of infrared radiation does take place from hot surfaces but it is not the only method and is probably, in many situations, the least important.

Cooling by Contact

Let us assume that the surface of the earth becomes heated in some way. (How this occurs will be dealt with in detail in a later chapter). Whenever an air molecule comes in contact by collision or conduction, the hot molecule of the surface will share some of its energy with the air molecule, with which it is in contact and so the surface cools down. This procedure occurs wherever the surface of the earth is hotter than the air molecules above it, which often happens over a very large part of the earth's surface and that includes the oceans as well as land areas.

Cooling by Evaporation

The oceans and all wet surfaces lose heat by a more important and effective procedure than simple contact. This is cooling by evaporation, which is important when one realises the extensive water surface of the earth. The oceans alone cover 70% of the earth's surface. This doesn't include all the lakes, rivers and surfaces that have become wet by rain. We, as human beings, are very much aware of this method of cooling when we wet the surface of our skin. Coming out of the bath always seems to be colder than being in the bath. The water on our skin is allowed to evaporate and so takes heat away from our body, which results in us feeling cold. This is how our bodies cool down naturally.

We call it sweating.

Cooling by evaporation takes place because the hotter or more energetic (faster) molecules in the water are more likely to leave the surface than do the slower and less energetic molecules. Water molecules are constantly moving in and out of the surface but, on the whole, the hot molecules have the advantage because they have enough kinetic energy to leave the surface. This simple procedure means that there is a net heat loss from the surface. Since most of the earth's surface is water, this method of losing heat is considerable and so there is this constant and enduring cooling by evaporation that takes place over the large wet parts of the earth. This ensures that there is plenty of water vapour and heat in the atmosphere.

Convection, Conduction and Diffusion

The heat that is carried by molecules in contact with the earth's surface does not remain near to the surface. This heat quickly moves away, which allows room for more heat to be transferred. The heat is dissipated in all possible ways of transferring heat. Radiation is accompanied by diffusion, convection and conduction. Let me remind you that conduction is the sharing of heat by contact or collision. The hot molecules transfer heat to the cooler molecules when they collide.

Convection and diffusion are the methods by which the heat is bodily carried away by the actual movement of the hot molecules. Diffusion is the spread of the molecules brought about by their random motion.

This is made more effective by having an atmosphere that is disturbed and turbulent. Wind has an effective cooling effect. Blowing over your hot cup of tea helps to cool it sufficiently to allow easy drinking.

We now have to consider another important factor that controls heat flow away from the earth's surface. This is the effect of the earth's gravitational field. It has two major influences, namely on convection and density. The density of the atmosphere is not uniform. The density gradually decreases with altitude. How does this reduction in density affect conduction? In a high density atmosphere, collisions between molecules are frequent, and as the density drops with altitude, then the number of collisions decreases.

Because collisions between molecules is the method by which the average kinetic energy (or temperature) is increased, then the temperature will gradually go down as altitude is gained where the frequency of collisions is diminished. We notice this by climbing a mountain and realising that warm clothing is required and that snow is more likely to appear and remain on the top of mountains. The temperature can reach really low values such as minus 50 degrees Celsius which is the temperature often experienced outside an airliner flying at 10 thousand metres, where the air density is very low.

Hot gases have a lower density than cold gases, which means that the gravitational field produces a force that makes the hot gases rise, leading to up-drafts and convection currents. So long as the hot gas has a higher temperature than the surrounding gases, heat transfer by convection will take place. We make use of this in the lifting of hot air balloons and in helping glider pilots to fly at high altitudes. So long as there is no obstruction to this flow of heat, then it will continue to rise until it reaches the very edge of space.

When the heat arrives at these high altitudes, sharing of energy becomes negligible because of the sparsity of molecules. This leaves radiation as the paramount energy at these high altitudes and consequently is the only method by which the earth loses its heat to the vacuum of outer space. Viewed from space, the earth is a ball of infrared radiation.

Chapter Six
How Does Earth's Surface Heat Up?

Let us assume that the earth's surface warms up in some way, then we are forced to ask the question, "From where does the heat energy come which subsequently warms up the surface?" We know that when an object heats up, such as the water in a hot water jug, we have no difficulty in recognising the source of the heat energy, namely electrical energy in the case of the hot water jug. Recognising the sources of the heat that heats up the earth's surface is only slightly more complicated. There are two sources of energy arriving at the earth's surface. The most obvious source is the sun that heats up the surface by emitting electromagnetic radiation which travels from the sun, taking eight minutes to reach the earth's surface.

The Hot Earth

But there is a second significant source and that is the earth itself. This latter source is often forgotten or ignored. It must be significant since the mass of the earth is made up almost entirely of a very hot liquid magma. Relatively speaking, the earth's solid crust is only a small part of the earth's mass and because of its low density, floats on the high density hot liquid. Although the heat leaving the hot internal earth may be less than the heat arriving from the sun, it would be wrong to ignore its contribution, especially in the oceans, which are mainly warmed up by conduction of heat through the earth's thin crust. Climate scientists, who are mostly concerned with what is happening in the atmosphere, tend to omit the earth as a source of energy probably because it is too difficult to estimate and measure the amount of heat involved. The El Nino effect, for instance, (heating of the surface of the ocean) is most likely caused by this geothermal energy.

The Bogus Greenhouse Effect

Much is made of the popular idea that the earth's surface is heated by the greenhouse effect that takes place in the atmosphere. A little consideration and thought will be enough to convince one that the earth's atmosphere is not a greenhouse. A greenhouse is a building designed to eliminate heat loss from its interior. It works on the principle that sunlight can enter, heats up the objects inside, which then emit infrared which cannot escape because the glass of the walls and roof is opaque to infrared radiation. The heat is reflected back into the greenhouse, and consequently there is a net input of heat energy into the greenhouse, leading to an increase in temperature. The greenhouse loses heat only by the slow conduction through the walls and roof. When the heat gained equals the heat lost, then an equilibrium temperature is reached. This equilibrium temperature depends on the rate at which heat enters the greenhouse and on the area and conducting properties of the walls and roof. If the temperature in the greenhouse reaches too high a level, then the competent greenhouse keeper opens a window, which is the quickest way to reduce the temperature. How does this happen? It happens because this allows the hot gases to escape by diffusion and convection from the inside through the open window to the outside.

But the earth's atmosphere is not a greenhouse; it is not designed to stop heat loss. A greenhouse has a solid roof or lid and walls, which are not only impervious to infrared but also stop any movement of warm gases passing to the outside. The solid walls and roof stop convection and diffusion from taking heat away and to some extent also stop the loss of heat by radiation. The atmosphere does not have a solid roof or walls. This needs to be emphasised because it is of paramount importance to the process of cooling. There is nothing to stop heat from escaping from the atmosphere. Heat in the lower atmosphere rises by convection and radiation, and moves freely into the upper atmosphere. Infrared heat radiation is almost the only type of energy present in the upper atmosphere, from where it is now free to escape to outer space. The earth is not a greenhouse because there is no solid roof to stop the heat loss. Furthermore, the atmosphere is subject to winds and turbulence, which help to transfer heat from warm areas to cold areas, enabling further cooling of the atmosphere.

The only situation that could possibly be described as greenhouse warming is when there is a thick cloud covering. In that case, the infrared radiation will be absorbed by the water droplets in the cloud and then re-emitted downwards. Upwards radiation will be reabsorbed by the water droplets and so the downward directed radiation is prominent.

However there are several points that have to be noted. First is that the absorption and re-emission is carried out by water; not carbon dioxide. Secondly, although cooling by radiation loss is eliminated, cooling by convection, diffusion and turbulence is not diminished.

Finally, cloud cover is not permanent and neither does it cover all the earth's surface. One day the sky may be completely cloud covered but the next day the sky may be completely clear. Besides that, because one place may be cloud covered doesn't mean that all places are cloud covered. Taking all these points into account, we can state quite definitely that cooling still exists even when there is thick cloud cover and most certainly, carbon dioxide plays no part in eliminating this cooling.

The Myth of Greenhouse Gases

The labelling of gases as being 'greenhouse gases' is wrong. It doesn't matter what type of gases are inside the greenhouse. It is not the gases that keep the heat inside. It is their containment by the walls and roof that makes the greenhouse work. The hot gases are not free to escape from inside an efficient greenhouse. If they are not contained, then they will spread naturally by diffusion and convection; they do not remain stationary. Because of this natural movement, gases cannot be described as being greenhouse gases. Gases cannot mimic a greenhouse. Greenhouse gases are a figment of one's imagination. There is no such thing as a greenhouse gas. It is a contradiction. When gases get hot in the atmosphere, they produce convection currents, where the gravitational effect on the low density hot gases sends them upwards to the upper atmosphere, where they finally lose heat by radiation. The only way in which gases can act like a greenhouse is if they form an impenetrable fixed layer. Such a fixed layer does not exist and cannot exist while we have a gravitational field and a turbulent atmosphere. The earth's atmosphere does not have a hard impenetrable roof.

It is hard to understand how this simple and obvious observation is ignored by those who keep on referring to the existence of the greenhouse effect. Because the earth's atmosphere is not a greenhouse, then one cannot use the theory of greenhouse entrapment to explain global warming, and, therefore, any predictions based on this erroneous theory must be suspect and not to be trusted.

The Atmosphere as Insulation

There is one other factor that needs to be considered when trying to understand how the surface of a planet can heat up and that is the insulating properties of the planet's atmosphere. The number of collisions between molecules and photons is most numerous near to the planet's surface where the density of the atmosphere is greatest; that is where the gravitational effect is greatest. When heat radiation is re-emitted, then it is likely, in this high density scenario, to hit another molecule and then be re-absorbed once more. The sharing of heat by collisions is also intense in this high density. At all times, the atmosphere at these low altitudes is full of energy, consisting of a chaotic mixture of kinetic and infrared energy, constantly interchanging from one sort to the other. Diffusion, convection and escaping radiation gradually allows the heat to rise through this chaotic activity until it reaches an altitude where collisions are less common. Then the heat radiation is free to escape into the cold of outer space.

This interchange of energy by collisions and re-emission of heat radiation is immense when the density is high and is consequently time consuming. These time-taking interactions delay the passage of heat upwards and so produce an effective insulating blanket. This results in maintaining a higher average temperature at levels nearer to the surface of the planet. This is fortunate for us on earth because without this insulating atmosphere, life on earth would be difficult. As these interactions diminish with height, we would expect these average temperatures to gradually decrease until they reach near temperatures of outer space. This lowering of temperature agrees with observations. In this situation, the only energy which is left is infrared radiation, which is now free of obstruction and can escape into the cold of outer space.

We can safely conclude that the insulating properties of an atmosphere depend mostly on its density and thickness. It does not at all depend on the type or size of the molecules. For example, the insulating properties of the atmosphere of Mars is very small resulting in very cold temperatures at the planet's surface. This is not surprising when one realises that the density and thickness of Mars's atmosphere is very low compared with that of the earth. The fact that Mars, which has an atmosphere that is almost entirely carbon dioxide, and with a carbon dioxide abundance twenty times greater than that on earth does not provide heat that one might expect to be entrapped by the greenhouse effect. Mars does not suffer from global warming.

In comparison, the atmosphere of Venus is very thick and dense and so one would expect the insulating properties of its atmosphere to be very effective. It is so effective that the heat arising from its hot interior has great difficulty in escaping. This leads to its surface temperature being very close to that of the hot magma that exists just below its surface. It is interesting to note that the amount of heat arriving at the surface of Venus from the sun is almost negligible dueto the density and thickness of its atmosphere. Most of the radiation arriving at Venus is reflected back into space caused by its very large albedo effect. This means that the hellish-like temperatures on its solid surface is not due to a run-away greenhouse effect, which is very commonly stated but simply due to its insulating efficiency of its dense atmosphere.

The fact that carbon dioxide is about one and half times more dense than air might lead one to believe that increasing the abundance of the more dense carbon dioxide by an extra two hundred parts per million would increase the density of the atmosphere sufficiently to increase its insulating properties by a sufficient amount enough to cause global warming. Adding another 200 ppm of carbon dioxide to the atmosphere would increase the atmosphere's density by about 0.03 %, which is a very small amount. To appreciate just how small this quantity is one must realise that when the temperature of the atmosphere changes by 10 degrees, its density changes by about 1%, which makes the carbon dioxide contribution to the earth's atmospheric density insignificant. This is why we have hot air balloons and not carbon dioxide free air balloons. Removing carbon dioxide from the air that fills balloons would not give sufficient buoyancy to raise the balloon off the ground.

Chapter Seven
What Causes Climate Change?

There is no doubt about the existence of climate change. The climate history of the earth provides us with enough evidence to reassure our belief that the earth's climate fluctuates on all scales of time and through small and large variations. This leads directly to the question, "What causes these numerous and various changes in climate?"

There are two possible ways in which the earth's atmosphere can change its temperature. The first one to consider is the change in the insulating properties of the atmosphere. We have discussed how this occurs and how it is dependent on the atmosphere's density and thickness. One might think that the insulation depends on the type of gases making up the atmosphere. For instance, the addition to the atmosphere of heavy gases that also have a large heat capacity might increase its insulating properties. To change the density, it would require a major influx of a high density gas. There is no evidence that there has been such a large influx and, consequently, no sudden increase in the density of the atmosphere during the last three million years. It might be suggested that carbon dioxide, with its large molecular weight, would be heavy enough to change the insulating properties of the atmosphere. This would require large quantities of the gas and there is no evidence of that ever occurring.

Carbon dioxide would also need to have a large heat capacity to entrap huge amounts of heat sufficient to change the climate. This cannot happen since the heat capacity of carbon dioxide is ten times smaller than water vapour and taking into account the greater abundance of water vapour, makes the heat holding properties of carbon dioxide less than significant. Even the heat capacity of oxygen is somewhat larger than that of carbon dioxide. With these considerations, we can safely say that changes in the carbon dioxide abundance is not a factor in the insulating properties of the atmosphere.

Changes in Heat Input

The second way in which the earth's atmosphere can change its temperature is very much more obvious. Ignoring changes to insulation, if we wish to raise the temperature of any object, such as the water in a kettle, then we simply apply more heat like switching on the electric heater. This is so obvious that one feels ashamed in mentioning it but, apparently, many climate scientists refuse to even consider this possibility. So long as the kettle's environment is lower in temperature, turning off the electricity will allow the kettle's temperature to fall by simple heat loss through radiation and conduction. Applying this idea to that of the atmosphere requires that we look for ways in which the heat input to the earth's surface can vary. In the cold of outer space, the earth's environment is certainly cold enough to ensure that the atmosphere will cool down as soon as there is a reduction in the heat input to it. The earth is constantly losing heat in the form of infrared radiation.

Because we have eliminated changes in the insulating properties of the atmosphere as a cause of climate change, we are left with just the one alternative and that is that climate change in the earth's atmosphere is produced by variations in the energy inputs to the earth's surface. So if we discover that the earth's atmosphere is heating up, then the question one must ask is "Where does the extra heat energy come from?" Then we must ask the additional question and that is, "Why and how does the heat input vary to produce a change in climate?"

In our attempt at answering these questions, we find that we have set ourselves a very difficult task. If we were able to supply answers, then we would be able to predict future climate changes. But we can't; when we attempt to do this, we find that it is almost impossible to make sensible predictions. The reason is because there are so many variables and factors controlling the amount of heat reaching the earth's surface. The study of heat flow into and out of the earth's surface is too complicated.

Changes in The Sun's Heat Input

Let's take the heat arriving at the earth's surface from the sun. This is practically entirely in the form of electromagnetic radiation. This is influenced by lots of factors which include many variations in the nature and functioning of the sun such as its wildly fluctuating magnetic field, the changes in its nuclear reactions and such things as its speed of rotation.

Other factors include alterations in the sun's surface temperature, unpredictable turbulence in the sun's interior and the uncertainty of its passage through the dust clouds that exist in the galaxy.

Even the earth's magnetic field is thought to have some effect on the heat arriving at the earth from the sun. Certainly variations in the earth's orbit (changing its distance from the sun) will influence the heat input. The nature of the earth's surface will also make a difference to the amount of heat it absorbs. Different surfaces have different reflecting powers. Cloud cover certainly affects the amount of heat reaching the earth's surface. The amount of cloud cover is itself unpredictable and, apparently, it is affected by the density of the cosmic rays hitting the earth's atmosphere. Water, forests, ice, snow, mountains and agriculture, all differ in their absorbing and reflecting abilities. Changes in ocean currents are very likely contributing factors in changing the heat input from the sun onto the earth's surface.

There are probably lots of other factors, known and unknown, affecting the heat input from the sun. We cannot be certain that we have included all the possible variables. If we wish to make sensible predictions, then not only must all these factors have to be taken into account but also how much each factor contributes to the change. This is an impossible task. We are not just looking for a single factor. We have to account for these factors acting in a chaotic combination. The overall result of all these possibilities acting to change the heat arriving at the earth's surface from the sun is that we cannot measure the total amount and certainly not enough to predict future climates.

The Fluctuating Heat Input from the Earth

The heat arriving from the hot, molten magma of the earth is also a heat source of unknown variation. Because the thickness and quality of the earth's crust varies so much, it makes it impossible to measure the heat transfer from the inside of the earth. The temperature, only a short distance down from the surface, is at such a high temperature, one would expect this heat flow to be significant, so significant that it cannot be ignored when considering climate change.

This heat flow from the earth is also dependent on many variables such as volcanic activity, thickness of the crust, movement of plate tectonics, conductivity of the earth's crust, radioactivity inside the earth, the convection currents in the magma and on the movement of water around the earth. We must remember that the crust is thinner under the oceans and so we would expect the oceans to be most effective in removing heat from the hot earth. We must be aware that there is far more volcanic activity under the oceans than on the solid surface of the earth, which is especially important since 70 percent of the earth's surface is ocean.

Unfortunately, just like the situation of the sun's variable input, there will be likely other variables, which affect the heat arriving at the earth's surface from the hot earth, which we just don't know about. Even if we can name all these factors affecting climate change, this isn't enough; we need to know by how much and under what conditions they act.

The overall situation is a case of chaos, where small changes can lead to big effects. All in all, it is an impossible task trying to predict the climate for any time in the future. It is a worthless and fruitless activity trying to predict future climates and stating that we can predict what the climate will be like in years to come is dishonest. This is wishful thinking but possibly and hopefully not a case of intended and hurtful deceit.

Chapter Eight
All About Carbon dioxide

In spite of all the evidence, advocates of anthropomorphic global warming continue to promote the ridiculous and incorrect idea that carbon dioxide (CO_2) is the agent causing climate change and continue to direct each of us to strive to reduce our carbon foot print. Because of this emphasis on making carbon dioxide appear to be a danger to our society, it is worthwhile examining some aspects and facts about carbon dioxide in order to eliminate some of the very wrong ideas that many people have.

Carbon Dioxide Is Not Poisonous

The first fact is that carbon dioxide is not poisonous. How do we, as ordinary individuals, know that carbon dioxide is not poisonous? The simple observation that we make this gas in our bodies and at any one time we have a large quantity in our lungs. We breathe out this gas and when we apply CPR, a high concentration of carbon dioxide is blown into our patients. It doesn't appear to do them any harm.

We consume quite a lot of carbon dioxide in various ways, such as when we drink fizzy soda water like beer and lemonade. Even when eating bread and cakes we take in a small amount of trapped carbon dioxide. It doesn't do us much harm except, perhaps, to cause us to burp.

Another occasion that reinforces the notion that carbon dioxide isn't poisonous is when we go partying in a room with poor ventilation. The carbon dioxide levels can reach quite high levels but they don't seem to do us much harm except perhaps to give us a headache. In this situation, it is not the presence of carbon dioxide that gives us a headache: it is the decrease in oxygen that causes the trouble. If we were trapped in an air-tight container, it is not the carbon dioxide that finally kills us, but the lack of oxygen. We can't live for long without oxygen but we can survive large concentrations of carbon dioxide so long as there is plenty of oxygen.

So why is it that many people believe carbon dioxide is poisonous? Maybe some people make the mistake of mixing up the names of carbon monoxide and carbon dioxide. Carbon monoxide is very poisonous. A more likely cause for the popularity of this incorrect idea about carbon dioxide is the misleading contributions of the media and the utterings of some so-called experts. There is also the possibility that people mix up smoke and carbon dioxide. This is often enhanced by the media when they show chimneys belching out smoke and steam. Carbon dioxide is a colourless gas and cannot be seen being emitted by chimneys. It is of interest to note that the technology nowadays allows us to remove the particles of matter from the emissions of modern power stations.

Carbon Dioxide Is a Fertiliser

Not only is carbon dioxide not poisonous but it is necessary for all life on the earth. Carbon dioxide is taken in by green plants and converted into sugars and oxygen by the process of photosynthesis. This is how we obtain our food. Without carbon dioxide in the atmosphere we would all die. We all need food. Because of photosynthesis, it is quite reasonable to consider this important gas as being a very necessary fertiliser. Indeed, if the level of carbon dioxide in the atmosphere were to fall below 200 parts per million (it is at present 400 ppm), there would be a disastrous fall in plant growth. Horticulturists are very much aware of the usefulness of carbon dioxide. They make their plants grow stronger and quicker by introducing high levels of carbon dioxide into their greenhouses. Those of us who are old enough will notice that our vegetables, including weeds, grow so much faster now the carbon dioxide levels are higher than they did long ago when the levels were lower. One should consider carbon dioxide our saviour and friend, and welcome more carbon dioxide into the atmosphere rather than trying to eliminate and demonise it.

Carbon Dioxide Is Not a Pollutant

Another misconception that is spread by the media and so-called experts is that carbon dioxide is a pollutant. A pollutant is some material introduced into our environment by industrialisation and by the general public, which is harmful to our health and well-being. It is true that industrialisation does introduce carbon dioxide into the atmosphere but it is not a danger to our health or environment. We have been given the impression that carbon dioxide is a new and harmful addition to the atmosphere. It is not new; it has been a major constituent in the earth's atmosphere since the earth's formation four billion years ago. It has always been with us since the dawn of time. Not only has it always been present in our atmosphere, the levels of concentration of carbon dioxide in these past ages have been much higher than now, sometimes hundreds of times greater. It should be worrying that the levels have been dropping since the beginning of time. This long term depletion is due to many living things such as sea creatures (like coral and shells) taking in CO_2 to produce limestone and chalk, which then gets locked away at the bottom of the oceans. Coal, peat and oil are other means by which carbon is removed and stored by living things.

It is of interest to note that we have a better case for calling oxygen a pollutant than carbon dioxide. Oxygen was not introduced into the atmosphere until the advent of green plants, some 3 billion years ago, a billion years after carbon dioxide. Oxygen was a newcomer to the atmosphere. When it was first introduced, it could easily have been described as a pollutant because much bacterial life at that time was poisoned by this very active gas.

Where Is the Earth's Carbon?

The rocks of the earth's crust containing carbon, such as limestone and fossil fuels contain much more carbon than exists in the oceans and atmosphere. While carbon dioxide is being replaced to the atmosphere by volcanism and the burning of fossil fuels, plant and animal life is removing the gas from the atmosphere. It is a race between removing and replacing; a race that in the past life of the earth has been won by removal.

The Carbon Dioxide and Temperature Correlation

One of the main arguments by the so-called climate scientists is that the graphs of carbon dioxide levels against time synchronise perfectly with the temperature-time graphs. There appears to be a strong correlation between the two measurements. High levels of CO_2 seem to fit with higher temperatures. The correlation isn't perfect; there have been times when the correlation was poor, but for the sake of the argument we will accept that there is a correlation.

We have to remember that correlation is not an argument for causation. Correlation can have three causes. Does event A cause event B or does event B cause event A or does event C cause both A and B? It is important to determine the order of these events. Does the increase in CO_2 levels come before the increase in temperature or does the increase in temperature come before the increase in CO_2?

It is not easy to decide on the order of these events just by observing the timing of these events. Normally one could simply look at what occurred first but in this case, it is difficult because of the technical problems in separating the times when the events are slow to show their existence and for their tendency to overlap. However, there have been attempts to determine which comes first. By close investigation of the carbon dioxide and temperature graphs, these investigators have concluded that there is some evidence which indicates that changes in temperature occur before the changes in carbon dioxide levels.

Because of the lack of certainty in measuring the timing aspect of these graphs, we have to use some other method of determining what came first. We can do this by studying the mechanisms by which one can cause the other. Studying just how one event can cause the other produces much more reliable evidence.

The amount of the carbon dioxide dissolved in the water of the oceans is very much greater than the amount in the atmosphere. One can describe the oceans as being a giant reservoir of carbon dioxide.

The amount carbon dioxide in water depends on two major factors, namely, pressure and temperature. The pressure in water of the oceans increases rapidly with depth and so its capacity to hold carbon dioxide increases immensely and therefore, any carbon dioxide released at depth has a good chance of being dissolved. We are easily reminded of this dependence on pressure when we open our bottle or can of soda water. Once we open the can, the gas pours out. The manufacturers of soda water use pressure to put carbon dioxide into the drink.

It is easy to understand how CO_2 levels increase with a temperature rise. A simple demonstration shows that water loses its dissolved CO_2 when it is heated and the oceans do have plenty of CO_2 to lose. The dependence on temperature is also easily demonstrated. One can remove the CO_2 from the soda water by the simple act of heating it up. The bubbles of CO_2 are readily observed bursting at the surface of the hot drink. There will always be an exchange of gas molecules going to and from the surface of the oceans. So if the oceans warm up due to heat rising from the hot earth, then the oceans will lose more carbon dioxide than they gain. If the temperature of the oceans falls, then the procedure is reversed. This to and from activity is easy to understand by the professional scientist as well as by any layman using just the simple observation of his own lungs, which emit the high concentration of carbon dioxide from his blood when it comes into contact with the low concentration of CO_2 in the air entering his lungs.

On the other hand, it is difficult to explain how an increase in CO_2 content of the atmosphere can cause an increase in temperature. Previous arguments in this monograph indicate quite clearly that it is very unlikely that carbon dioxide could cause climate change. There is not enough of the gas, and it doesn't have the capacity to store any quantity of heat.

When scientists have two explanations of an occurrence, they use Occam's razor to decide which is most likely to be the correct one. This states that the simpler explanation is likely to be the correct one. The simple comparison made here can only strengthen one's belief that a rise in temperature comes before a rise in CO_2.

Some argue that carbon dioxide cannot escape from the sea water because it is tied chemically to water. The argument states that the sea is acidic because of the absorption of carbon dioxide. It is true that carbon dioxide will react weakly with pure water to produce carbonic acid, but the oceans are simply not acidic. They are slightly alkaline because of their reaction with basic basalt rocks which cover most of the sea floors. The carbon dioxide existing in water has not made a chemical reaction, but exists in molecular form as pure CO_2. So the transfer of carbon dioxide between the atmosphere and the oceans is not impeded by any chemical reaction, and so can enter and leave easily with any temperature change.

Chapter Nine
Speculations and Perceptions

In this last chapter, I have chosen to deviate from my philosophy of writing about evidence and nothing but evidence. I apologise to my readers for not being constant in this endeavour. After presenting strong evidential and obvious reasons against the man-made theory for climate change, I realised that many readers would feel somewhat let down and uncomfortable being left with the puzzling question as to why this erroneous theory is believed by so many people. Since I am not a sociologist or psychologist (my expertise is in the physical sciences), I can only make speculations about the cause for this aberration. Nonetheless, I think that it would be interesting and worthwhile to know some of the ideas that are prevalent in the debate on climate change and I hope that the rumours, opinions, propositions, suggestions and perceptions presented here may stimulate the reader to consider why such a ridiculous philosophy has invaded and settled in the minds of so many followers.

But before you continue to read this, I must warn you that if you want the truth about climate change, you must not be influenced by these opinions, even if they appear valid and are from people who are considered to be authorities and leaders. There is only one way to determine the truth about climate change, and that is by studying the relevant scientific and historical evidence.

Intergovernmental Panel on Climate Change

The Intergovernmental Panel on Climate Change (IPCC) is the main organisation promoting the idea of anthropomorphic global warming (AGW). I became suspicious of the proclamations of the IPCC when I realised that they were not a scientific organisation, even though they have stated that they consist of scientists. They are in fact a political group attached to the United Nations. Their title tells us that its members are government appointees. Unfortunately, politicians and government agencies are not well-known for their adherence to the truth. For instance, the IPCC has claimed and still claims that most scientists agree with their main claim that man-made carbon dioxide produces global warming by means of the greenhouse effect. Their statement that 97% of all scientists agree with the anthromorphic global warming theory (AGW) is ridiculous. It would be impossible to get that number of scientists to agree on any subject. They seem to ignore the 50,000 scientists who signed a petition against AGW and the many reputable scientists who have written articles and books decrying AGW. I have elsewhere listed names of a hundred sceptical authors, all of whom are experts in their fields. I think that if I included in my list all those scientists who have not voiced or written about their scepticism, the numbers would be ten times greater.

Certainly, there are lots of scientists who have criticised the IPCC for its adherence to AGW.

Predictions of the Intergovernmental Panel on Climate Change

The anthropomorphic global warming theory can never be disproved because no matter what severe weather event occurs, the adherents of this theory are keen to point out that they predicted the event in the first place. Whether the event is a cooling trend or a heat wave, whether it is a drought or a flood, according to the AGW advocates, the AGW theory predicted it. This is an example of 'post-diction' and not prediction. In other words, AGW is not science, it is more like a religion which depends on faith rather than evidence.

Instead of an angry god producing bad weather, according to the AGW advocates, it is you, by your production of carbon dioxide, that causes severe weather events. For example, the infamous 'hockey stick curve', which was quoted by the IPCC and by Al Gore in his film (The Inconvenient Truth), was finally shown to be a case of nothing less than fraud, after a courageous and persistent attempt by independent researchers to obtain the relevant data. This 'hockey stick curve' is a graph of temperatures over the last thousand years or so but which showed that the temperature of the earth had been very nearly constant over that period of time. Well, it hasn't been constant over historical times. The graph had eliminated the Medieval Warming period and the Little Ice Age, but showed a dramatic increase in temperatures towards the end of the 1990s. The steepness of the curve was frightening, which is certainly what the author meant it to be.

This exponential rise in temperature did not materialise, and so the prediction of a deadly rise in temperature failed.

Using Computers

Computers can take large quantities of measurements and convert them, according to a program, into a final result or average. No matter how quick or how large the computer is, the final result is dependent on two factors. There is the reliability of the input measurements and there is the dependability of the program put into the computer to calculate the final result from the input measurements. We know that temperature measurements are particularly unreliable, and the programs used to find the average are man-made and, because of that, are subject to not only human error, but also to human bias.

The program is a model or theory chosen by the computer users. Like all theories, there is considerable doubt about the worthiness or validity of the programs or models used in the computers. In all, the adage 'garbage in gives garbage out' is only too true when using computers. Using computers to predict a future event is, therefore, very dependent on the information input and on the model entered into the computer. The statement that computers only tell you what you want to hear is unfortunately too often true. This must be considered to be a likely possibility with those favouring AGW.

The Bias of the IPCC

Scientific work, before being published, should be 'peer reviewed', which is supposed to discourage poor science by being reviewed by other scientists in the same field of work.

Unfortunately the system is easily abused by selecting reviewers who have ulterior reasons for denying or accepting work for publishing. The IPCC claim that their work is always peer reviewed but their reviewers are chosen from a small group who already believe in AGW. Independent reviewers are not sought and are positively discouraged. The existence of the Medieval Warm period is an obvious damaging fact to the AGW believers. This was made public when certain emails between AGW adherents were leaked. These emails sent between AGW scientists indicate the existence of a serious non-scientific agenda. The 'climate-gate' emails showed conclusively that these so-called scientists were very concerned with making the Medieval Warm period disappear. This ignoring of the history of climate change and the attempts at eliminating past warm periods is a damning situation for the followers of AGW, since the production of CO_2 by industrialisation was not available in these early periods. This means that in all probability the present warm period is natural and not man-made; a conclusion that is completely opposite to the statements and belief of the IPCC.

IPCC Predictions

The predictions of the adherents of AGW have mostly turned out to be erroneous. Capable scientists would automatically discard their theory as being unreliable and false when their predictions failed to occur. But the AGW people simply go on making more predictions and constantly stating that we are approaching a tipping point, which is supposed to be a point after which we have no hope of recovering. The unfortunate situation for the AGW theory is that there have been tipping points predicted for every year since the year 2000. These tipping point predictions have ranged from days to years in the near future. For example, a UK prime minister, said in 2009 that we only had 50 days left. Recently a presidential candidate in US has said that we are 10 years too late to stop a climate catastrophe. Even a royal prince in 2009 stated that we had only 96 months left to save the earth from disaster. Tipping points have been announced by pop stars, TV personalities and other celebrities. It goes without saying that none of these tipping points have occurred. These examples of 'Cry Wolf' don't appear to deter the AGW adherents from making these unsubstantiated predictions of climate disaster, but they have since learnt that it is advisable and safer to predict events far into the future, 50 to 100 years ahead, when most of us will not be around to determine their validity.

Does the IPCC Tell Lies?

In their propaganda, the IPCC have fallen to telling lies about severe weather events. The data is quite clear about this; severe weather events are no more frequent than they were in the past. This is effective propaganda because severe weather events are frightening, and fear is a useful weapon for persuasion, panic thinking and unreasonable actions. If severe weather events don't frighten you into believing in AGW, then the plight of the drowning polar bears will surely convince you of AGW. It is certain that the IPCC were aware of the fact that polar bears are thriving and are actually increasing in numbers.

There is an implied lie when the IPCC use the greenhouse theory because they must be aware that the earth is not a greenhouse, in that the earth does not have a roof or walls as does a real greenhouse. This is an example of where, if you say it often enough, then someone will believe it no matter how ridiculous it is.

Changing the Objective of IPCC

Another sequence of events which made me suspicious of the IPCC was the changing of their goal. At first, they were trying to convince us that the greenhouse effect was warming the earth, but since the greenhouse effect was easily discounted by anyone who knows that the earth doesn't have a roof, they changed their slogan proclaiming that global warming was paramount. For the first decade of this century, global warming wasn't happening, so the IPCC dropped that idea. Nowadays, the IPCC have limited their aims to announcing that there is climate change, and everyone is to blame for it. No one can argue that climate doesn't change; sceptics and believers all agree that it does, and so the IPCC can feel quite safe that they cannot be easily criticised for claiming that climate changes, since it actually does. Nonetheless, they haven't dropped the idea that you cause climate change by your wasteful life style and, consequently, their aim to make you feel guilty because of your production of carbon dioxide has not gone away.

The Closing of Debate

Debate and criticism is the basis of true science. The AGW adherents are not keen on debating the issue of global warming. Their reactions range from downright refusal to allow the opposite point of view, such as denying any possibility of publishing letters to the editor in local newspapers, and by issuing statements such as, "It is too late for debate." In my talks to local groups, I consistently ask my audiences to tell me where I have gone wrong in my analysis. In my letter writing to newspapers and to AGW advocates, I also plead for anyone to inform me of my mistakes. So far, I have not had any reply other than abuse. This obscurantism too often leads to personal abuse of sceptics, describing them as deniers, zombies, flat earthers, traitors, holocaust deniers, idiots, ignoramuses and persons requiring psychological help. Too many discourage debate by threatening sceptics with all sorts of punishments ranging from sackings to imprisonment and even execution. Censorship of this intensity usually means that there is something to hide; a fear that the truth will out, showing the extent of the fraud. If the AGW adherents have full confidence in their theory, then they should be only too pleased to have the chance to prove their case. Climate sceptics are often accused of being in the pay of the big oil corporations, I can assure my readers that I am not receiving any financial aid from anyone.

AGW Celebrities

Besides the members of the IPCC, there are many well-known celebrities who have been drawn in to support AGW. The first one that came to my knowledge is Al Gore who was a vice-president of the USA. He made a film called *The Inconvenient Truth* which is a brilliant propaganda film using all the clever techniques meant to stir the emotions. It presented the hockey-stick graph, frightening us with temperatures rising exponentially. It pictured melting ice of glaciers and starving polar bears stranded on small icebergs. It was so obviously constructed to frighten people that a judge in the UK felt that it was necessary to forbid it being shown in British schools.

Gore advises us to stop producing CO_2 by changing our wasteful and indulgent lives but he personally has a very lavish life style, where he jets frequently around the world, ironically, promoting his film. He has several mansions, each using as much electricity as 30 average American homes. One of these is built on the beach which is surprising considering that he warns us about catastrophic sea level rising. He is obviously not awfully worried about the sea flooding his home.

A famous film star, who is a vocal AGW advocate, has been made the United Nations messenger of peace. He travels in a private jet around the world doing what he describes as 'good work' but demands that you and I have to travel less and use buses. Has he ever been on a bus? Besides the multiple homes he owns, he has a luxury yacht which, no doubt, uses fossil fuels and so is not carbon free. This is the person who described CO_2 as a poison. There are many other high flyers, who have lavish life styles belying their promotion of a simpler life style. All these advocates of AGW tell you to change your habits, but are unwilling to change their own lives.

Isn't this extreme hypocrisy?

Why Do So Many People Believe in AGW?

With all the obvious evidence from science and history, it is hard to understand why so many people believe that climate change is man-made and, in particular, by the production and emission of carbon dioxide. What is it that drives the man-made climate change activists? There is no simple answer and, I admit, that I do not know the reason why this hysteria about climate has spread so effectively around the world. There have been other scientific frauds, but this one is more endemic, and has infected more ordinary people than any other scam.

There are lots of suggestions that have been made by others to try and explain why this theory has been so successful. The obvious one is financial gain. Governments are quite generous in supplying grant money for research on climate change and quite keen to extract extra taxes in the form of carbon credits. At least one AGW advocate has been made into a millionaire in the market of carbon credits. Much of the tax money has gone to corporations who are ready to take advantage of this easy cash in supplying carbon free power generators and transport.

Another suggestion is that some people feel important when predicting the end of the world, and so egotism is one of the strong motivations for proclaiming AGW and the catastrophic end of civilisation. There are numerous AGW advocates who have predicted tipping points leading to a descent into a hot hell-like world.

Unfortunately, another motivation is simply the hatred of mankind. The idea that mankind causes everything that is bad in the world is a popular feeling. There are many people who believe that we all should be punished for being human. It needs very little imagination to see the similarity between this hatred and the religious belief in original sin. Carbon dioxide is a very convenient object for marking us appear sinful, because we all produce carbon dioxide, even if we do nothing else but breathe.

United Nations Organisation and AGW

Another suggested motivation for advocating AGW is that it enables some to have power over others. Because CO_2 emissions are worldwide, this provides an ideal vehicle for providing international control and so is attractive to the United Nations Organisation, which has whole heartedly accepted the idea of AGW. There is no doubt that inducing fear in the population has been used in the past by tyrannical autocracies to enable them to control their peoples, and there is no reason why we should not believe that this objective is being used by the United Nations and present governments.

Genuine Believers

Of course many adherents of AGW may be motivated by a genuine conviction that carbon dioxide emissions are causing global warming. This is understandable when they are subject to very effective propaganda and lies. Most people are not scientists, and have no ideas about the scientific method, or about applying uncertainties in measurements.

History is also a dark area for many people; comparing events of the now and present, with that of the past, is an academic procedure not often applied by ordinary people, and is left to so-called experts.

Most people are concerned with their ordinary everyday activities, and, according to Bertrand Russell, "Most people would rather die than think." We cannot criticise these people because nearly everyone wants to do what is right and save the earth. This is understandable when they are told constantly by governments and the media that they are guilty of destroying their world.

Most people are decent and honest and because of this, they are likely to believe that everybody else is also decent and honest. It is hard to believe that people in authority do often tell lies, and so these honest people tend to believe what they are told. These decent people are not only being fooled but are also made to feel guilty unnecessarily, and this I find despicable.